W9-AOU-298

A SECOND POETRY BOOK

compiled by John Foster

illustrated by
Alan Curless
Paddy Mounter
Martin White
Joe Wright

Oxford University Press 1980

Oxford University Press, Walton Street, Oxford OX2 6DP

Oxford London Glasgow
New York Toronto Melbourne Wellington
Nairobi Dar es Salaam Cape Town
Kuala Lumpur Singapore Hong Kong Tokyo
Delhi Bombay Calcutta Madras Karachi

ISBN 0 19 918137 3 (non net)
ISBN 0 19 918136 5 (net)

Typeset by Tradespools Ltd, Frome
Printed in Hong Kong

Contents

Why?

Why are the leaves always green, Dad?
Why are there thorns on a rose?
Why do you want my neck clean, Dad?
Why do hairs grow from your nose?

Why can dogs hear what we can't, Dad?
Why has the engine just stalled?
Why are you rude about Aunt, Dad?
Why are you going all bald?

Why is Mum taller than you, Dad?
Why can't the dog stand the cat?
Why's Grandma got a moustache, Dad?
Why are you growing more fat?

Why don't you answer my questions?
You used to; you don't any more.
Why? Tell me why. Tell me why, Dad?
Do you think I am being a bore?

John Kitching

Me

My mum is on a diet,
My dad is on the booze,
My gran's out playing Bingo
And she was born to lose.

My brother's stripped his motorbike
Although it's bound to rain.
My sister's playing Elton John
Over and over again.

What a dim old family!
What a dreary lot!
Sometimes I think that I'm the only
Superstar they've got.

Kit Wright

Juster and Waiter

My mum had nicknames for me and my brother.
One of us she called Waiter
and the other she called Juster.
It started like this:
she'd say, 'Lend me a hand with the washing up
will you, you two?'
and I'd say, 'Just a minute, Mum.'
and my brother'd say
'Wait a minute, Mum.'
'There you go again' – she'd say,
'Juster and Waiter.'

Michael Rosen

My Dad, Your Dad

My dad's fatter than your dad,
Yes, my dad's fatter than yours:
If he eats any more he won't fit in the house,
He'll have to live out of doors.

Yes, but my dad's balder than your dad,
My dad's balder, OK?
He's only got two hairs left on his head
And both are turning grey.

Ah, but my dad's thicker than your dad,
My dad's thicker, all right,
He has to look at his watch to see
If it's noon or the middle of the night.

Yes, but my dad's more boring than your dad.
If he ever starts counting sheep
When he can't get to sleep at night, he finds
It's the sheep that go to sleep.

But my dad doesn't mind your dad.
Mine quite likes yours too.
I suppose they don't always think much of US!
That's true, I suppose, that's true.

Kit Wright

Shut that door!

Leaving a door ajar
Is something my Mother can't bear,
'Were you born in a cave?' she asks.
Well, she should know,
She was there!

Julie Holder

Mother's Nerves

My mother said, 'If just once more
I hear you slam that old screen door,
I'll tear my hair out! I'll dive in the stove!'
I gave it a bang and in she dove.

X. J. Kennedy

If you don't put your shoes on

If you don't put your shoes on before I count
fifteen then we won't go to the woods to climb
the chestnut tree one

    But I can't find them

Two

    I can't

They're under the sofa three

    No
    Oh yes

Four five six

    Stop – they've got knots they've got knots

You should untie the laces when you take your shoes
off seven

    Will you do one shoe while I do the
    other then?

Eight but that would be cheating

    Please

All right

    It always . . .

Nine

    It always sticks – I'll use my teeth

Ten

    It won't it won't
    It has – look

Eleven

    I'm not wearing any socks

Twelve

    Stop counting stop counting. Mum where
    are my socks mum?

They're in your shoes. Where you left them.

    I didn't

Thirteen

    Oh they're inside out and upside down and
    bundled up

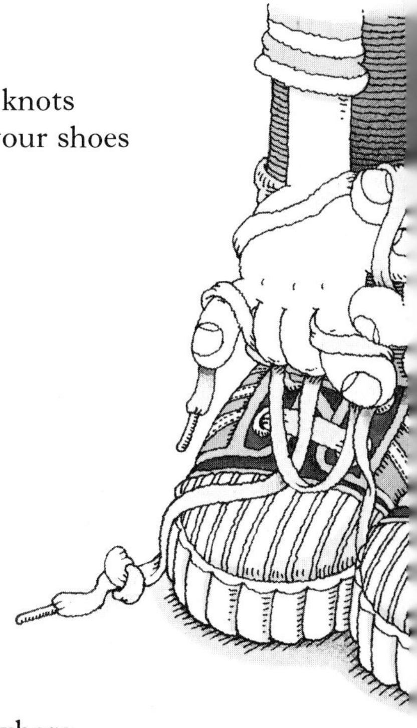

Fourteen
        Have you done the knot on the shoe you
        were . . .
Yes
Put it on the right foot
        But socks don't have right and wrong foot
The shoes silly
Fourteen and a half
        I am I am – wait.
        Don't go to the woods without me
        Look that's one shoe already
Fourteen and threequarters
        There
You haven't tied the bows yet
        We could do them on the way there
No we won't fourteen and seven eighths
        Help me then
        You know I'm not fast at bows
Fourteen and fifteen sixteeeenths
        A single bow is all right isn't it?
Fifteen we're off
        See I did it.
        Didn't I?

Michael Rosen

Orders of the Day

Get up!
Get washed!
Eat your breakfast!
That's my mum,
Going on and on and on and on and on . . .

Sit down!
Shut up!
Get on with your work!
That's my teacher,
Going on and on and on and on and on . . .

Come here!
Give me that!
Go away!
That's my big sister,
Going on and on and on and on and on . . .

Get off!
Stop it!
Carry me!
That's my little sister,
Going on and on and on and on and on . . .

Boss
Boss
Boss
They do it all day.
Sometimes I think I'll run away,
But I don't know
Where to go.

The only one who doesn't do it,
Is my old gran.
She says,
'Would you like to get washed?'
Or,
'Would you like to sit on this chair?'
And she listens to what I say.

People say she spoils me,
And that she's old-fashioned.
I think it's the others that spoil;
Spoil every day.
And I wish more people were old-fashioned,
. . . like my gran.

John Cunliffe

Gran

'My goodness! What a big boy you are!
Good gracious! How you've grown!'
The first words always from my gran:
They always make me moan.

'You must have been drinking your milk up;
You must have been eating your greens;
You must have been going to bed early.'
I'm really fed up with these scenes.

Do you think I should tell her
How small she is? 'I do believe you've shrunk.
Why are you growing a beard, Gran?
You're beginning to look like a monk.'

'You must have been drinking the gin, Gran.
You eat too much pudding and cake.
You've been watching the Hulk on TV, Gran.
How about a quick jump in the lake?'

But no. I'll just grin it and bear it
And take growing pains like a man.
I'll kiss her moustache; hold my hand out –
– And get pocket money from Gran.

John Kitching

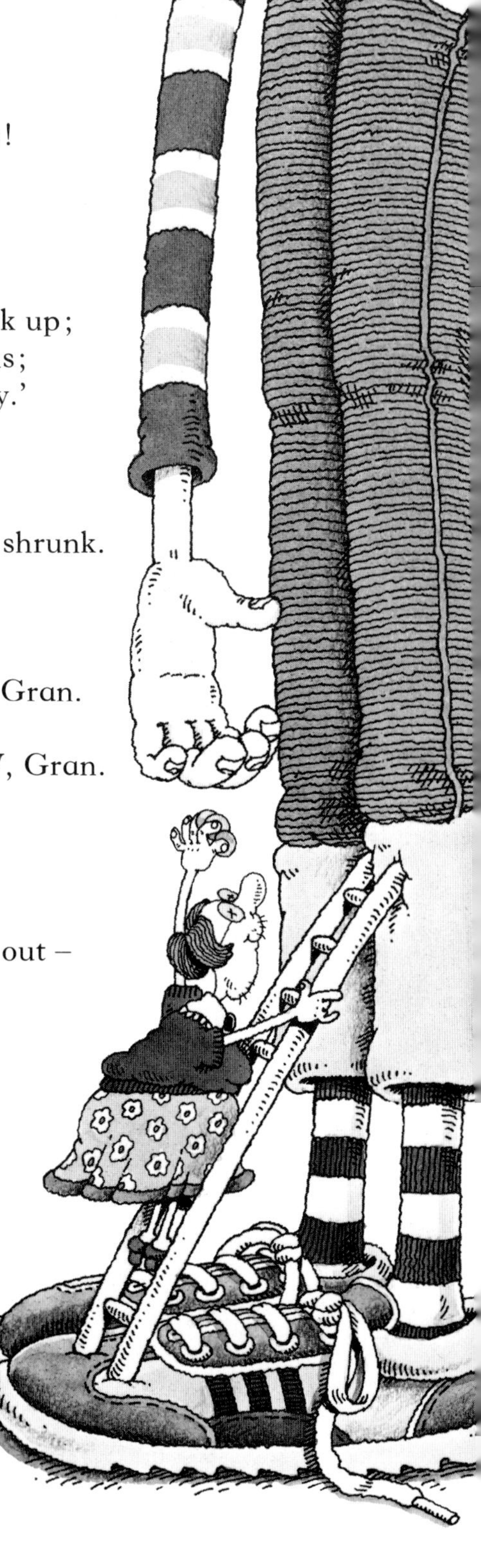

My Gramp

My gramp has got a medal.
On the front there is a runner.
On the back it says:
Senior Boys 100 Yards
First William Green
I asked him about it,
but before he could reply
Gran said, 'Don't listen to his tales.
The only running he ever did
was after the girls.'
Gramp gave a chuckle
and went out the back
to get the tea.
As he shuffled down the passage
with his back bent,
I tried to imagine him,
legs flying, chest out,
breasting the tape.
But I couldn't.

Derek Stuart

Legging the Tunnel

I don't know whether I believe
All Grandpa says:
Whether there ever really was
A tunnel
From side to side
Only an inch or so more wide
Than a canal boat;
Or that
Wrapped in a boatman's duffel coat
He had a ride
Under the Wren's Nest Hill.

The water, he said, lay black and still
When they cast off the horse,
And nosed the boat by pole
Into that dark, unwinking hole.

I can remember Grandpa's words:
'Out of a world of sky and trees and birds
We slid;
Out of the yellow sun
Into the rifled barrel of a gun
It seemed;
No path,
Only a tube
Of brickwork half awash with water
Where a shout
Bounced off the rounded, echoing roof
Ridged with worn bands
Of masonry;
Ribbed to the touch of groping hands
Like a knitted sleeve turned inside out.

'Two miles of tunnel
If you snuffed the lantern half-way through
And lost
The capering ogres on the walls
The blackness sucked and blew
In a whistle past you to a pin-hole of light,
The tunnel end.
You were alone
In a pipe of starless night;
Only the slap and chuckle of the water on the sides
As the boatman lay
Shouldering back
On the cabin ramp
And walked the ceiling like a fly
With hob-nailed boots.

'We were
Legging the tunnel.'

Gregory Harrison

Cousin Reggie

Cousin Reggie
who adores the sea
lives in the Midlands
unfortunately.

He surfs down escalators
in department stores
and swims the High Street
on all of his fours.

Sunbathes on the pavement
paddles in the gutter
(I think our Reggie's
a bit of a nutter).

Roger McGough

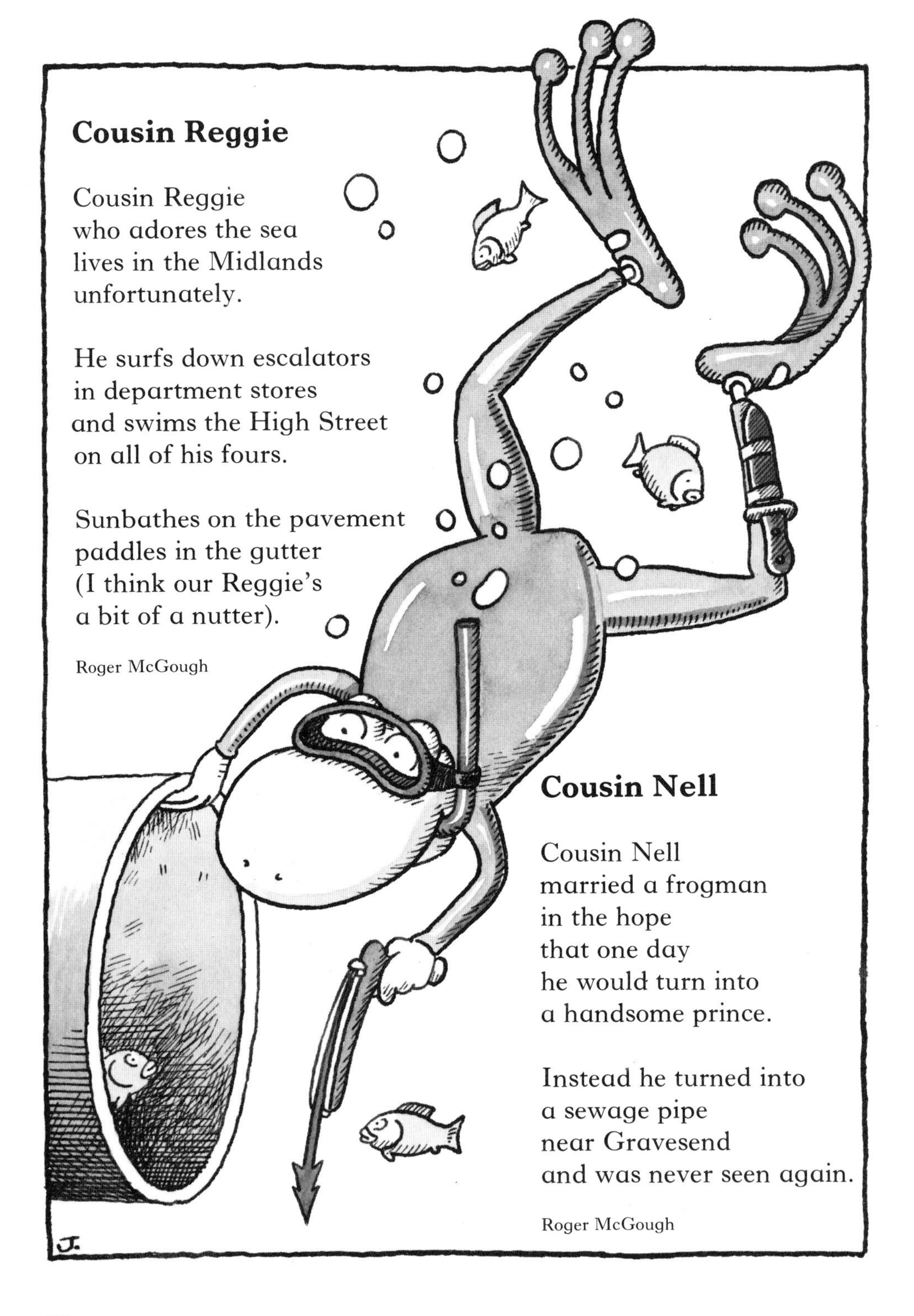

Cousin Nell

Cousin Nell
married a frogman
in the hope
that one day
he would turn into
a handsome prince.

Instead he turned into
a sewage pipe
near Gravesend
and was never seen again.

Roger McGough

Night Starvation or The Biter Bit

At night, my Uncle Rufus
(Or so I've heard it said)
Would put his teeth into a glass
Of water by his bed.

At three o'clock one morning
He woke up with a cough,
And as he reached out for his teeth –
They bit his hand right off.

Carey Blyton

The Longest Journey in The World

'Last one into bed
has to switch out the light.'
It's just the same every night.
There's a race.
I'm ripping off my trousers and shirt,
he's kicking off his shoes and socks.

'My sleeve's stuck.'
'This button's too big for its button-hole.'
'Have you hidden my pyjamas?'
'Keep your hands off mine.'

If you win
you get where it's safe
before the darkness comes –
but if you lose
if you're last
you know what you've got coming up is
the journey from the light switch to your bed.
It's the Longest Journey in the World.

'You're last tonight,' my brother says.
And he's right.

There is nowhere so dark
as that room in the moment
after I've switched out the light.

There is nowhere so full of dangerous things,
things that love dark places,
things that breathe only when you breathe
and hold their breath when I hold mine.

So I have to say:
'I'm not scared.'
That face, grinning in the pattern on the wall,
isn't a face –
'I'm not scared.'
That prickle on the back of my neck
is only the label on my pyjama jacket –
'I'm not scared.'
That moaning-moaning is nothing
but water in a pipe –
'I'm not scared.'

Everything's going to be just fine
as soon as I get into that bed of mine.
Such a terrible shame
it's always the same
it takes so long
it takes so long
it takes so long
to get there.

From the light switch
to my bed
it's the Longest Journey in the World.

Michael Rosen

Midnight Wood

Dark in the wood the shadows stir:
   What do you see? –
Mist and moonlight, star and cloud,
Hunchback shapes that creep and crowd
   From tree to tree.

Dark in the wood a thin wind calls:
   What do you hear? –
Frond and fern and clutching grass
Snigger at you as you pass,
   Whispering fear.

Dark in the wood a river flows:
   What does it hide? –
Otter, water-rat, old tin can,
Bones of fish and bones of a man
   Drift in its tide.

Dark in the wood the owlets shriek:
   What do they cry? –
Choose between the wood and river;
Who comes here is lost forever,
   And must die!

Raymond Wilson

The Malfeasance

It was a dark, dank, dreadful night
And while millions were abed
The Malfeasance bestirred itself
And raised its ugly head.

The leaves dropped quietly in the night,
In the sky Orion shone;
The Malfeasance bestirred itself
Then crawled around till dawn.

Taller than a chimney stack,
More massive than a church,
It slithered to the city
With a purpose and a lurch.

Squelch, squelch, the scaly feet
Flapped along the roads;
Nothing like it had been seen
Since a recent fall of toads.

Bullets bounced off the beast,
Aircraft made it grin;
Its open mouth made an eerie sound
Uglier than sin.

Still it floundered forwards,
Still the city reeled;
There was panic on the pavements,
Even policemen squealed.

Then suddenly someone suggested
(As the beast had done no harm)
It would be kinder to show it kindness,
Better to stop the alarm.

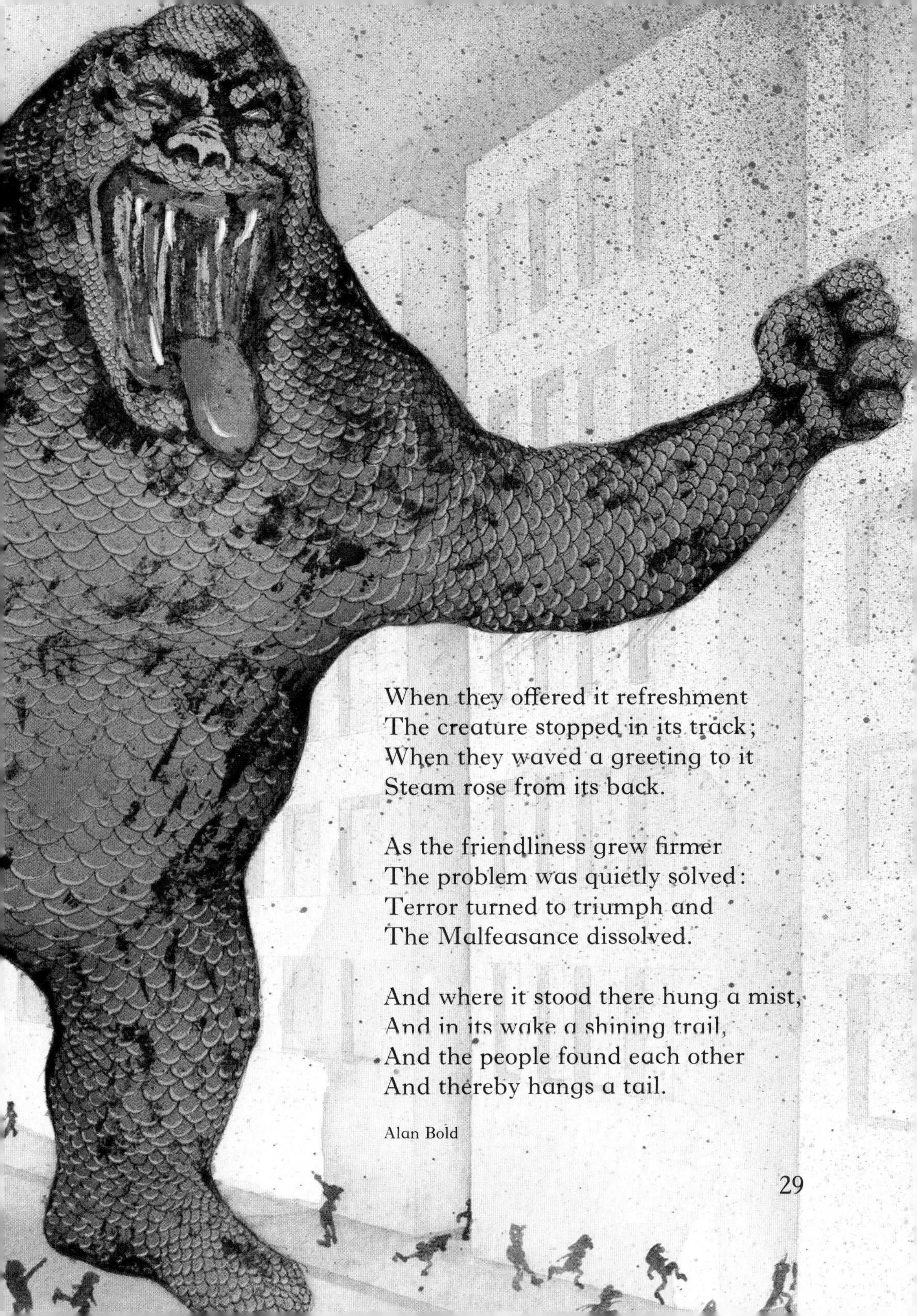

When they offered it refreshment
The creature stopped in its track;
When they waved a greeting to it
Steam rose from its back.

As the friendliness grew firmer
The problem was quietly solved:
Terror turned to triumph and
The Malfeasance dissolved.

And where it stood there hung a mist,
And in its wake a shining trail,
And the people found each other
And thereby hangs a tail.

Alan Bold

The Visitor

A crumbling churchyard, the sea and the moon;
The waves had gouged out grave and bone;
A man was walking, late and alone . . .

He saw a skeleton on the ground;
A ring on a bony finger he found.

He ran home to his wife and gave her the ring.
'Oh, where did you get it?' He said not a thing.

'It's the loveliest ring in the world,' she said,
As it glowed on her finger. They slipped off to bed.

At midnight they woke. In the dark outside,
'Give me my ring!' a chill voice cried.

'What was that, William? What did it say?'
'Don't worry, my dear. It'll soon go away.'

'I'm coming!' A skeleton opened the door.
'Give me my ring!' It was crossing the floor.

'What was that, William? What did it say?'
'Don't worry, my dear. It'll soon go away.'

'I'm reaching you now! I'm climbing the bed.'
The wife pulled the sheet right over her head.

It was torn from her grasp and tossed in the air:
'I'll drag you out of bed by the hair!'

'What was that, William? What did it say?'
'Throw the ring through the window! THROW IT AWAY!'

She threw it. The skeleton leapt from the sill,
And into the night it clattered downhill,
Fainter . . . and fainter . . . Then all was still.

Ian Serraillier

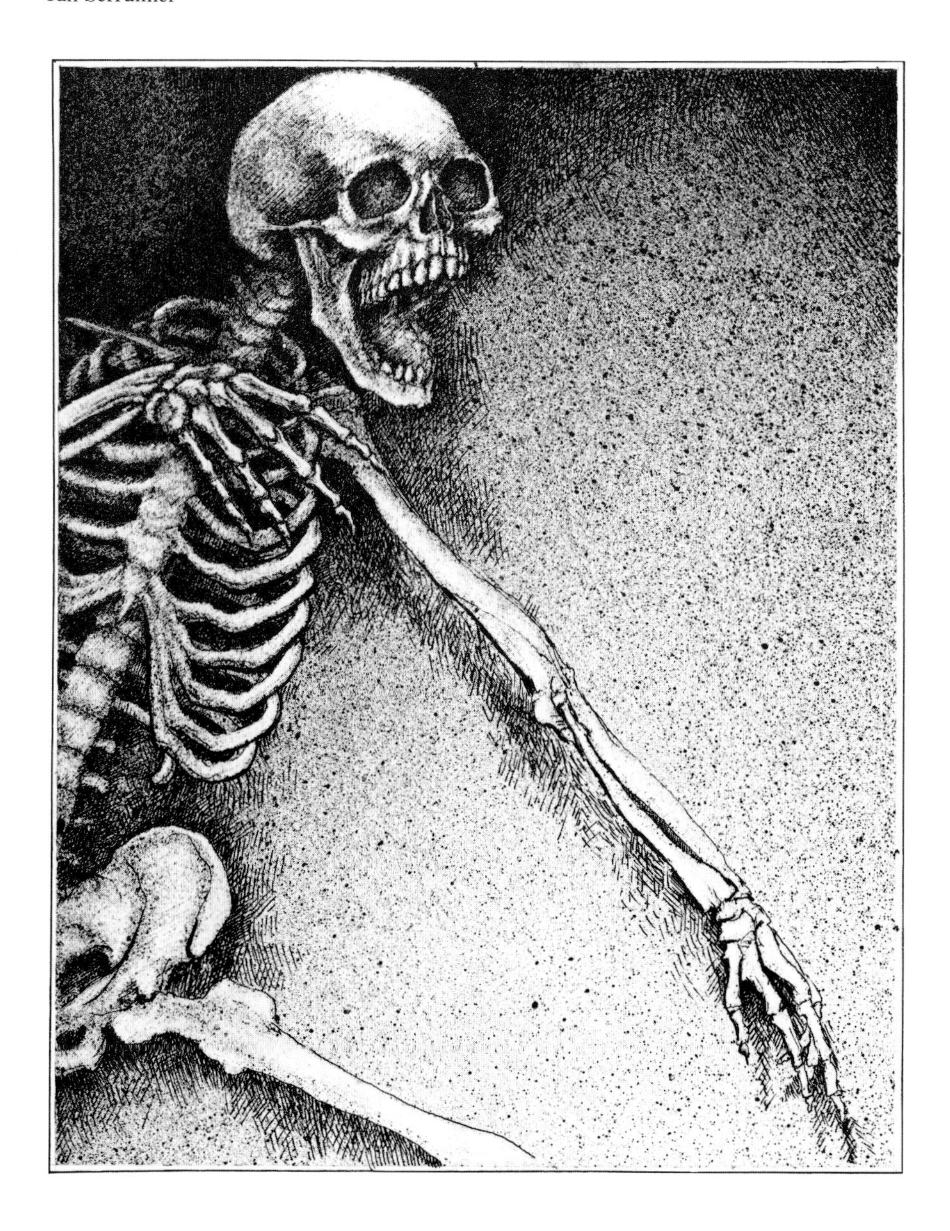

The Ogre

In a foul and filthy cavern
where the sun has never shone,
the one-eyed ogre calmly gnaws
a cold and moldy bone.

He sits in silence in the slime
that fills his fetid home
and notes the nearing footsteps
in the monstrous catacomb.

The one-eyed ogre drools with joy,
his stony heart beats fast,
he knows that for some girl or boy
this day shall be their last.

He wields his ugly cudgel
in a wide and vicious arc,
it swiftly finds his victim
in the deep and deadly dark.

Then down and down and down again
the ogre's blows descend,
to rend, and render senseless,
to speed his victim's end.

So pity those who stumble through
the one-eyed ogre's cave –
that dark abode he calls his home
shall surely be their grave.

Jack Prelutsky

Giant Denny

Giant Denny's clothes
have seen better days.
His shoes are scuffed.
There is a button missing
from his waistcoat.
His tie has soup stains on it.
His fingers are bony and gnarled,
like the branches of an oak.
His skin is rough like bark.
When he reaches down
to pinch my cheek,
I am not frightened,
only sad.

Derek Stuart

The Ghoul

The gruesome ghoul, the grisly ghoul,
without the slightest noise
waits patiently beside the school
to feast on girls and boys.

He lunges fiercely through the air
as they come out to play,
then grabs a couple by the hair
and drags them far away.

He cracks their bones and snaps their backs
and squeezes out their lungs,
he chews their thumbs like candy snacks
and pulls apart their tongues.

He slices their stomachs and bites their hearts
and tears their flesh to shreds,
he swallows their toes like toasted tarts
and gobbles down their heads.

Fingers, elbows, hands and knees
and arms and legs and feet –
he eats them with delight and ease,
for every part's a treat.

And when the gruesome, grisly ghoul
has nothing left to chew,
he hurries to another school
and waits . . . perhaps for you.

Jack Prelutsky

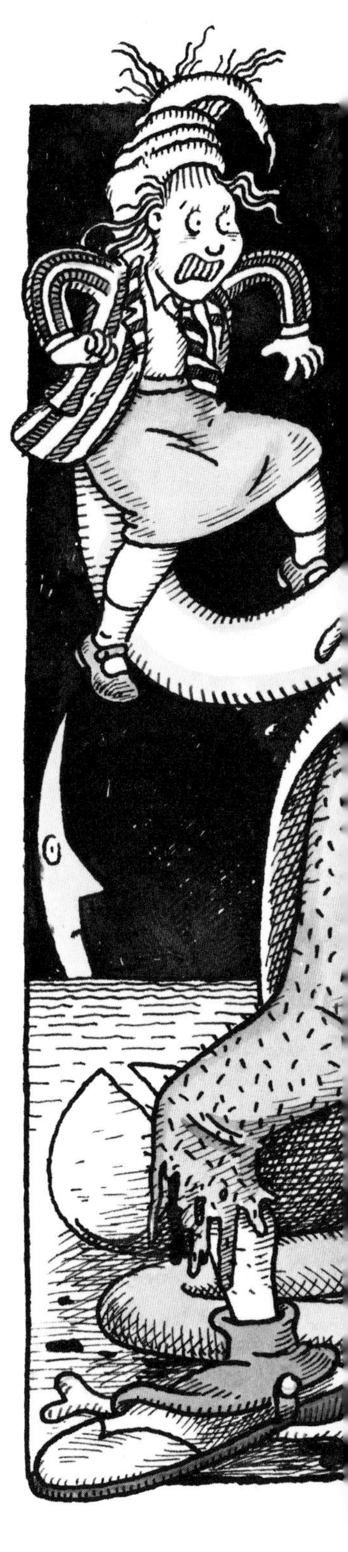

'Be a monster'

I am a frightful monster,
My face is cabbage green
And even with my mouth shut
My teeth can still be seen.
My finger-nails are like rats' tails
And very far from clean.

I cannot speak a language
But make a wailing sound.
It could be any corner
You find me coming round.
Then, arms outspread and eyeballs red,
I skim across the ground.

The girls scream out and scatter
From this girl-eating bat.
I usually catch a small one
Because her legs are fat;
Or it may be she's tricked by me
Wearing her grandpa's hat.

Roy Fuller

The Plug-hole Man

I know you're down there, Plug-hole Man,
In the dark so utter,
For when I let the water out
I hear you gasp and splutter.

And though I peer and peek and pry
I've never seen you yet;
(I know you're down there, Plug-hole Man,
In your home so wet.)

But you will not be there for long
For I've a *plan*, you see;
I'm going to catch you, Plug-hole Man,
And Christian's helping me.

We'll fill the bath with water hot,
Then give the plug a heave;
Then rush down to the outside drain –
And *catch* you as you leave.

Carey Blyton

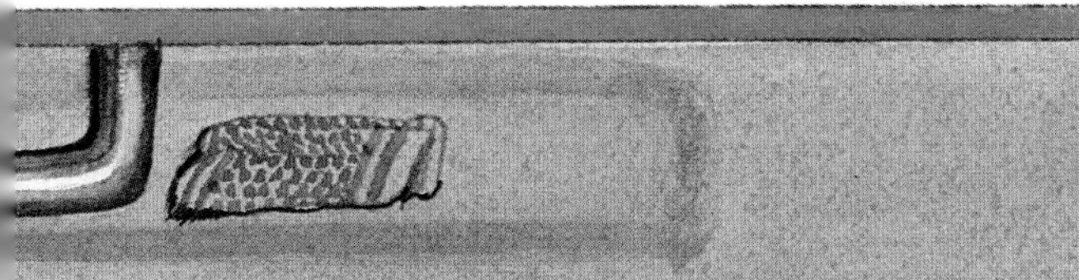

In the Bathroom

What is that blood-stained thing –
Hairy, as if it were frayed –
Stretching itself along
The slippery bath's steep side?

I approach it, ready to kill,
Or run away, aghast;
And find I have to deal
With a used elastoplast.

Roy Fuller

The Spangled Pandemonium

The spangled pandemonium
Is missing from the zoo
He bent the bars the barest bit,
And slithered glibly through.

He crawled across the moated wall,
He climbed the mango tree,
And when the keeper scrambled up,
He nipped him in the knee.

To all of you, a warning
Not to wander after dark,
Or if you must, make very sure
You stay out of the park.

For the spangled pandemonium
Is missing from the zoo,
And since he nipped his keeper,
He would just as soon nip you!

Palmer Brown

Surprise! or The Escapologist

At the foot of the Apennine Mountains,
Where the astronauts explore,
The curious lunar sea-squirt
Watches from its door.

And as they wander blithely
Around in their lunar cars,
The sea-squirt steals their module –
And flies away to Mars.

Carey Blyton

Action Man

Polar explorer
Action Man
Covers his ears with fur
In his bright red hood
And puts his ski-ing boots on.
As he bends his knees
And crouches over his skis
It begins to freeze,
Snowy mountains rise
Into frosty skies
And over the snow
He swoops towards the Pole.

Action Man puts on jump-boots
And parachute.
As he fastens his helmet
Under his chin
Clouds gather round
Looking like the ceiling
Of the roof of the world
Where they live
To people below –
And, pulling the ripcord,
Through the ceiling
The Red Devil dives.

Action Man puts on his silver suit
And heavily weighted boots.
As he puts his head
Inside his helmet
The countdown has started;
There is his rocket
Pointed like a pencil,
Ready to dot
A landing spot
On the moon.
Beneath it the fuel flares
And like an extra star
It shoots through the air.

The winds give a lion-like roar
And huge waves somersault
Like elephants onto the shore
As eagle-eyed Action Man
Puts his equipment on
And with his crew
Speeds to the rescue
In a motor boat
Or lowers a rope
From a twirling, whirling
Helicopter
While savage waves beneath
Snap their hungry teeth;
But nothing can
Frighten Action Man.

Stanley Cook

Why?

Superman can fly.
   Why can't I?
Popeye can swim,
   But I'm not him.
Paddington's a bear –
   It's not fair!

Why should all the people on TV
Have so much more fun than me?

Why can Mr Bounce tumble?
Why can Miss Piggy sing?
Why can a Womble womble?
When I can't do anything?

If you didn't know the answer,
   you'd want to cry.
I do know the answer. They're not real.
   I am. That's why.

Gyles Brandreth

Doctor Who

Doctor Who, I am forced to admit,
Is a booby, a crackpot, a twit.
  Other Time Lords all laugh
  When he trips on his scarf
That took X plus Y light years to knit.

Charles Connell

You tell me

Here are the football results:
League Division Fun
Manchester United won, Manchester City lost.
Crystal Palace 2, Buckingham Palace 1
Millwall Leeds nowhere
Wolves 8 A cheese roll and had a cup of tea 2
Aldershot 3, Buffalo Bill shot 2
Evertonill, Liverpool's not very well either
Newcastle's Heaven, Sunderland's a very nice place 2
Ipswhich one? You tell me.

Michael Rosen

One Tuesday when I was about ten

One Tuesday when I was about ten
I swam thirty-two lengths
which is one mile
and when I climbed out of the water
I felt like a big fat lump of jelly
and my legs were made of rubber
and there was this huge man there
with tremendous muscles all over him
and I went up to him and I said,
'I've just swum a mile.'
and he said,
'How many lengths was that, then?'
'Thirty-two' I said,
and the man looked into the water and said,
'I've got a lad in here who can do ninety.'

Michael Rosen

Elastic Jones

Elastic Jones had rubber bones.
He could bounce up and down like a ball.
When he was six, one of his tricks
Was jumping a ten-foot wall.

As the years went by, Elastic would try
To jump higher, and higher, and higher.
He amazed people by jumping a steeple,
Though he scratched his behind on the spire!

But, like many a star, he went too far,
Getting carried away with his power.
He boasted one day, ‘Get out of my way,
I’m going to jump Blackpool Tower.’

He took off from near the end of the pier,
But he slipped and crashed into the top.
Amid cries and groans, Elastic Jones
Fell into the sea with a plop.

Derek Stuart

Until I saw the sea

Until I saw the sea
I did not know
that wind
could wrinkle water so.

I never knew
that sun
could splinter a whole sea of blue.

Nor
did I know before
a sea breathes in and out
upon a shore.

Lilian Moore

A Flock of Little Boats

A flock of little boats
Tethered to the shore
Drifts in still water . . .
Prows dip, nibbling.

Samuel Menashe

Old Man Ocean

Old Man Ocean, how do you pound
Smooth glass, rough stones round?
Time and the tide and the wild waves rolling
Night and the wind and the long grey dawn.

Old Man Ocean, what do you tell,
What do you sing in the empty shell?
Fog and the storm and the long bell tolling,
Bones in the deep and the brave men gone.

Russell Hoban

Six White Skeletons

Deep deep down in the sea in the deep sea darkness
where the big fish
flicker and loom
and the weeds are alive
like hair

the hull of the wreck
grates in the sand;
in and out
of its ribs of steel –
only the long eel
moves there.

Down in the engine-room
six white skeletons;

only the long eel
moves there.

Kit Wright

A Baby Sardine

A baby sardine
Saw her first submarine;
She was scared and watched through a peephole.

'Oh, come, come, come,'
Said the sardine's mum,
'It's only a tin full of people.'

Spike Milligan

A sea-serpent saw a big tanker

A sea-serpent saw a big tanker,
Bit a hole in her side and then sank her.
    It swallowed the crew
    In a minute or two,
And then picked its teeth with the anchor.

Anon.

The Donkey Boys

I pass them on the seashore early,
Two Spanish boys in the sun. One ten,
The other six. They wave and smile at me,
Then bend their ragged backs again

To search the driftwood and sort out
Light dry sticks for bedding. Their donkey
Is small and white. He stands there without
Moving. His eyes and ears are sleepy.

The panniers that they must heap
Up high are huge. They work steadily,
Smiling as they go. As if asleep,
Their donkey follows quietly.

One morning I met them riding
Past the school where children were at play.
The boys didn't look at them, but smiling
Went on gently to their work that day.

Albert Rowe

In the Playground

In the playground
Some run round
Chasing a ball
Or chasing each other;
Some pretend to be
Someone on TV;
Some walk
And talk,
Some stand
On their hands
Against the wall
And some do nothing at all.

Stanley Cook

Upside Down

It's funny how beetles
and creatures like that
can walk upside down
as well as walk flat;

They crawl on a ceiling
and climb on a wall
without any practice
or trouble at all,

While I have been trying
for a year (maybe more)
and still I can't stand
with my head on the floor.

Aileen Fisher

A Wisp of a Wasp

I'm a wisp of a wasp with a worry,
I'm hiding somewhere in Surrey
I've just bit upon
The fat sit upon
Of the King – so I left in a hurry!

Colin West

Don't spray the fly, Dad

Don't spray the fly, Dad.
I like to see its wings
Shuddering and shaking.
Busily it sings.

Busily and buzzily
It walks across the glass,
And bumps against the blind
Which will not let it pass.

Up and down it prances;
Up and down it dances.
It hops and it leaps.
It never, never sleeps.

I like to see it caper.
Don't whack it with the paper.
It isn't very kind, Dad.
It isn't very kind.

John Kitching

Way down south where bananas grow

Way down south where bananas grow,
A grasshopper stepped on an elephant's toe.
The elephant said, with tears in his eyes,
'Pick on somebody your own size.'

Anon.

The Flea and the Ox

Flea

Why are you
(So large and strong)
A drudge for man
A slave so long?

I drink his blood,
On his skin I lurk;
I never do
A whit of work.

Ox

My master's good and feeds me well;
He made the house in which I dwell.
I haul his cart, I plough his track;
And when he's pleased he pats my back.

Flea

Oh, I couldn't have that,
For I'm not hefty;
Pat my back,
And-what'd-be-left-o'me?

Ian Serraillier

The Spider

The spider fishes the air with its net
Where the flies and insects go past;
Its web is fine as lines of a fingerprint
Or scratches on a pane of glass.

Round as a ball of string,
With silken ladders stretching
Out from the centre where it is sitting,
It walks on air in every direction.

It stitches up a hole
In the empty air
Or hangs on its own rope,
When its web is broken, to make a repair.

A spider is a miniature monster
That draws its web on a shadow
High up in a dusty corner
Or comes in the house and curtains the window.

Outside on a misty morning the trap it spreads
Catches dew instead of flies
And the sun picks pearls from the threads
As they gradually dry.

And in the winter the scores of webs
Have nothing left to catch
In the leafless hedge
But the breath of the frost.

Stanley Cook

Centipede's Song

'I've eaten many strange and scrumptious dishes
in my time,
Like jellied gnats and dandyprats and earwigs
cooked in slime,
And mice with rice – they're really nice
When roasted in their prime.
(But don't forget to sprinkle them with just a pinch
of grime.)

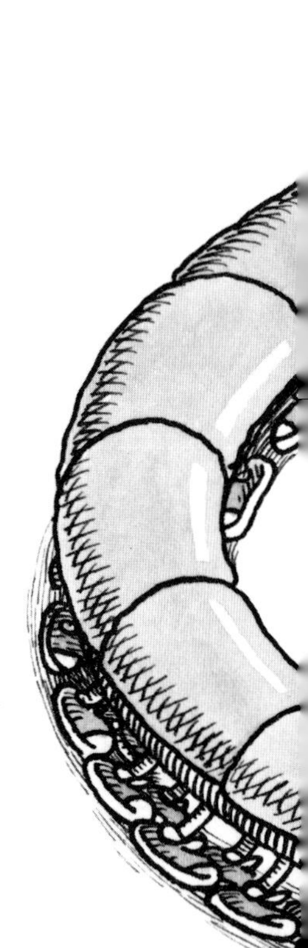

'I've eaten fresh mudburgers by the greatest cooks
there are,
And scrambled dregs and stinkbug's eggs and
hornets stewed in tar,
And pails of snails and lizards' tails,
And beetles by the jar.
(A beetle is improved by just a splash of vinegar.)

'I often eat boiled slobbages. They're grand when
served beside
Minced doodlebugs and curried slugs. And have
you ever tried
Mosquitoes' toes and wampfish roes
Most delicately fried?
(The only trouble is they disagree with my inside.)

'I'm mad for crispy wasp-stings on a piece of
buttered toast,
And pickled spines of porcupines. And then a
gorgeous roast
Of dragon's flesh, well hung, not fresh –
It costs a pound at most,
(And comes to you in barrels if you order it by post.)

'I crave the tasty tentacles of octopi for tea
I like hot-dogs, I LOVE hot-frogs, and surely
you'll agree
A plate of soil with engine oil's
A super recipe
(I hardly need to mention that it's practically free.)

'For dinner on my birthday shall I tell you what I
chose:
Hot noodles made from poodles on a slice of
garden hose –
And a rather smelly jelly
Made of armadillo's toes.
(The jelly is delicious, but you have to hold your nose.)

'Now comes,' *the Centipede declared*, 'the burden
of my speech:
These foods are rare beyond compare – some are
right out of reach;
But there's no doubt I'd go without
A million plates of each
For one small mite,
One tiny bite
Of this FANTASTIC PEACH!'

Roald Dahl

Two Sad

It's such a shock, I almost screech,
When I find a worm inside my peach!
But then, what *really* makes me blue
Is to find a worm who's bit in two!

William Cole

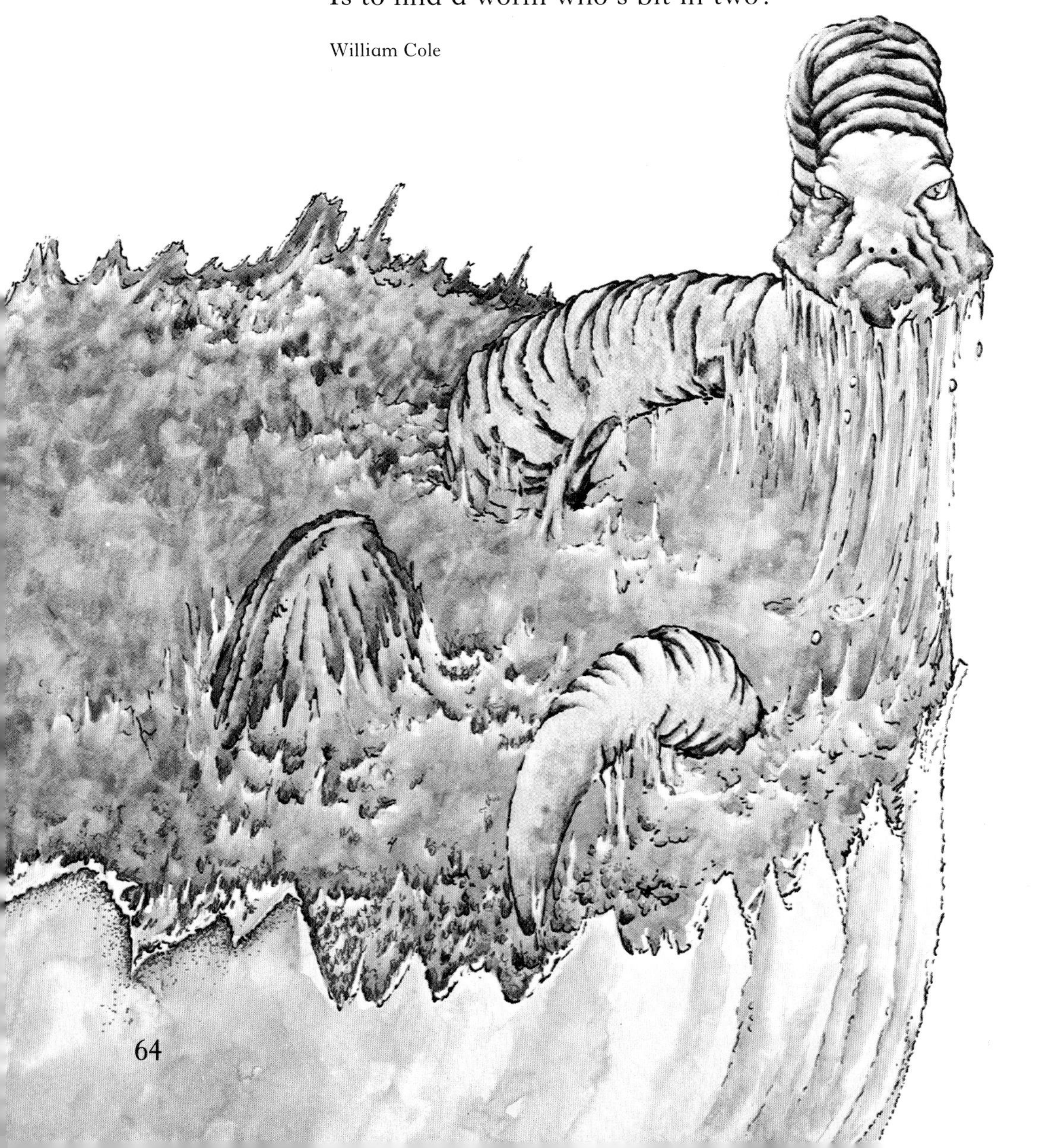

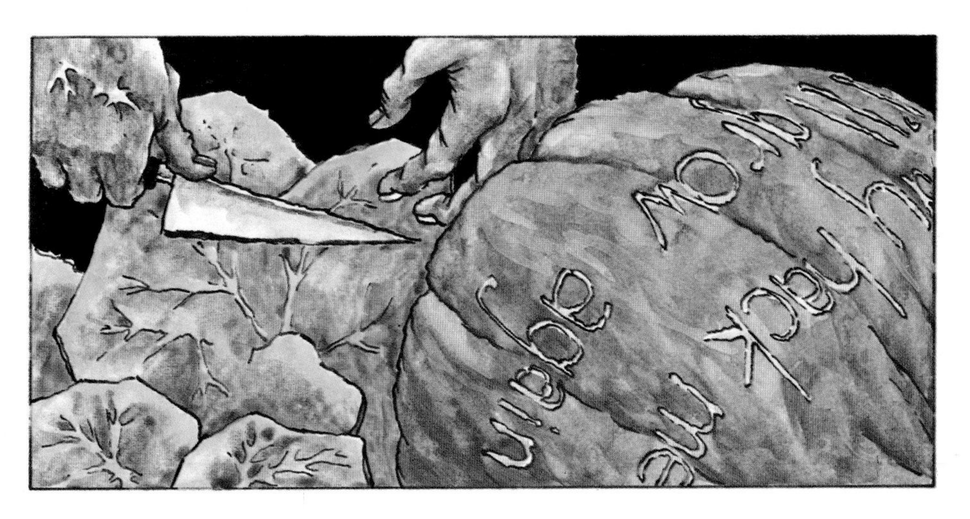

The Pumpkin

You may not believe it, for hardly could I:
I was cutting a pumpkin to put in a pie,
And on it was written in letters most plain
'You may hack me in slices, but I'll grow again.'

I seized it and sliced it and made no mistake
As, with dough rounded over it, I put it to bake;
But soon in the garden as I chanced to walk,
Why there was that pumpkin entire on his stalk!

Robert Graves

The Toaster

A silver-scaled dragon with jaws flaming red
Sits at my elbow and toasts my bread.
I hand him fat slices, and then, one by one,
He hands them back when he sees they are done.

William Jay Smith

End of a Girl's First Tooth

Once she'd a tooth that wiggled;
Now she's a gap that lisps.
For weeks she could only suck lollies;
Now she champs peanuts and crithsps.

Roy Fuller

Chips

Out of the paper bag
Comes the hot breath of the chips
And I shall blow on them
To stop them burning my lips.

Before I leave the counter
The woman shakes
Raindrops of vinegar on them
And salty snowflakes.

Outside the frosty pavements
Are slippery as a slide
But the chips and I
Are warm inside.

Stanley Cook

A Hungry Old Goat Named Heather

A hungry old goat named Heather,
Was tied up with an old bit of leather.
In a minute or two
She had chewed it right through,
And that was the end of her tether.

Celia McMaster

The Hippopotamus

The huge hippopotamus hasn't a hair
on the back of his wrinkly hide;
he carries the bulk of his prominent hulk
rather loosely assembled inside.

The huge hippopotamus lives without care
at a slow philosophical pace,
as he wades in the mud with a thump and a thud
and a permanent grin on his face.

Jack Prelutsky

Black Bear

The bear's black bulk
Is solid sulk

He mopes with his nose
Between his toes

Or rears with a roar
Like a crashed-back door
Shouting: 'God swore

I'd be Adam –
And here I am

In Paradise.'
But then his eyes

Go pink with rage:
'But I am in a cage

Of rough black hair!
I have to wear

These dungfork hands!
And who understands

The words I shout
Through this fanged snout?

I am God's laugh!
I am God's clown!'
Then he glooms off
And lays him down

With a flea in his ear
To sleep till next year.

Ted Hughes

The Musk-ox

The musk-ox is the weathercock
Up on the world's snowladen roof.
The stars crowd shivering down to share
The central heating of his hoof.

And solid seas, sixty below,
Huddle about his spirit's blaze.
And hills of everlasting snow
Crouch at the peephole of his gaze.

Winds that have clanged around the pole
From the dawn of the world till now
Hammer the iron of his thoughts
Upon the anvil of his brow.

Ted Hughes

Town Fox

I used to see them cantering over the hill,
Across open fields burning to autumn's brown,
Hounds yelping madly, mouths watering for the kill,
Twenty scarlet horsemen bobbing up and down

As they chased the scared fox, riding side by side,
Horn rasping and loud shouts raising the alarm,
Leaping over hedges and walls in that fresh countryside,
Then ambling back at sundown to gabled manor and farm.

The fox is still hunted through the soft-echoing air,
Is flushed out from the cover of his stinking den
To race with cunning to some place of safety where
He is out of sight for a while from cruel dogs and men,

Turning his sharp nose to the scent on every small breeze
Fanning over the grasses that would give him away,
Then moves in the silence to a thick clump of trees
As rain sprinkles down to save him from dying that day.

Now he is leaving his parish of gorse, heather and wheat,
Is seen slinking along dark pavements all hours of the night,
Padding down avenues, crossing the busiest street,
Killed often by cars, eyes glowing green in their light.

Lives in the ruins of the dead squire's town house,
Raising the cubs where wine bottles were once stored,
Sharing the sun on the wild garden with spider and mouse,
Sleeping all day in a jumble of mortar and board.

Not threatened here, and is glad to be left alone,
To roam where he will, raid all the neighbouring bins,
Cabbage stalks, kipper skins, the surprise of a chicken bone;
We have almost forgiven the fox for his so-called sins.

So justice at last in his rough, peaceful domain
Where once the guests danced, the chandeliers glowed,
He has outlived them all, come into his own again,
Has forgotten the hounds, how hard those old hunters rode.

Leonard Clark

The Canary

The song of canaries
Never varies,
And when they're molting
They're pretty revolting.

Ogden Nash

Alley Cat

A bit of jungle
    in the street
He goes on velvet toes,
And slinking through the shadows
    stalks
Imaginary foes.

Esther Valck Georges

Ode to a Goldfish

O
Wet
Pet!

Gyles Brandreth

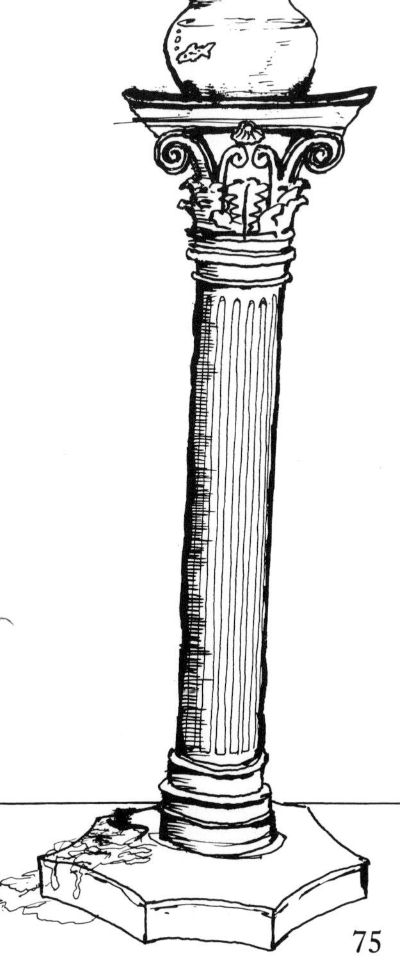

My Gerbil

Once I had a gerbil –
Bought me by my Dad
I used to watch it in its cage,
Running round like mad
Or sleeping in a corner
Nesting in a hole
Made of shavings, bits of wool
And chewed up toilet roll.

I kept it in the kitchen
In the cage my cousin made.
It flicked all bits out on the floor
Mum grumbled – but it stayed.
I fed it; gave it water;
Was going to buy a wheel.
I used to take it out sometimes –
To stroke. I liked the feel –
All soft, with needle eyes,
A little throbbing chest.
I'd had a bird, a hamster too:
The gerbil I liked best.

I came downstairs one morning.
I always came down first.
In the cage there was no movement.

At once I knew the worst.
He lay there in the corner.
He'd never once been ill –
But now, fur frozen, spiky,
No throbbing, eye quite still.

I tell you – I just stood there
And quietly cried and cried,
And, when my Mum and Dad came down,
I said, 'My gerbil's died'.

And still I kept on crying,
Cried all the way to school,
But soon stopped when I got there
They'd all call me a fool.

I dawdled home that evening.
There, waiting, was my mother.
Said: 'Would you like another one?'
But I'll never want another.

John Kitching

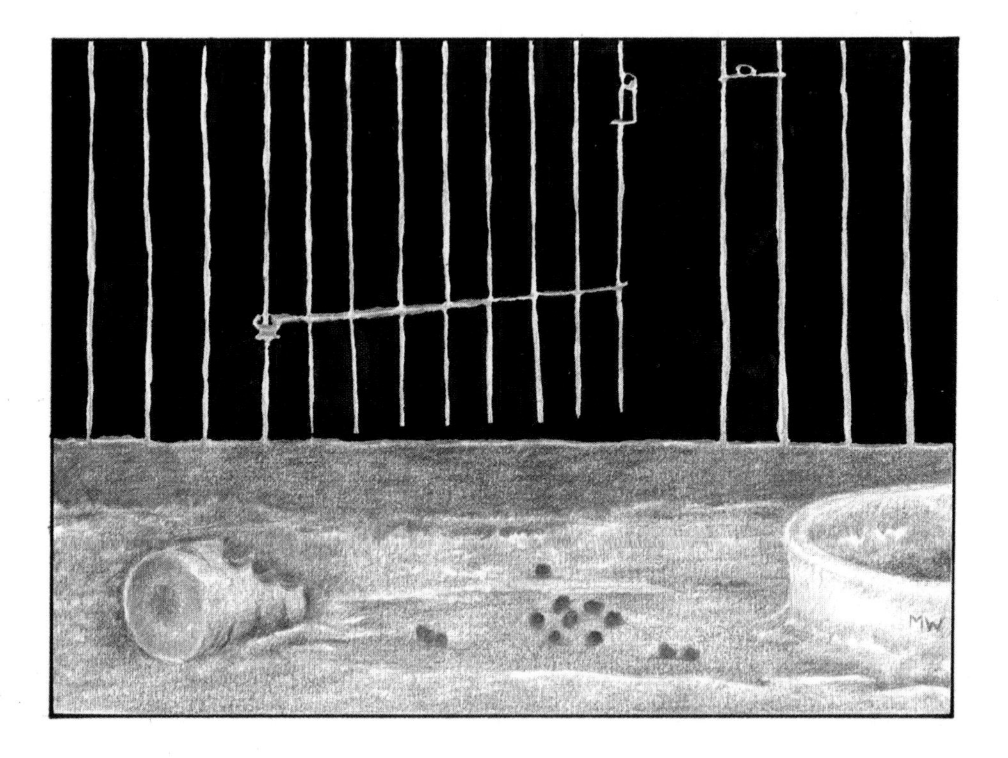

Four Fish

I Pike

Still as a stone
This waterlogged log
Lies all alone
Digesting a frog;

But see how the shoals
Of small-fry flinch
As he tilts his scales
At a pound to the inch.

II Rudd

Swimming as close to the sun as they are able
They turn the afternoon into a fable,
Rolling like golden coins on a miser's table.

III Tench

This underwater navvy with the blood-shot eye
Is a hard-shovelling mud-lover, black as green can be;
When he's earned his grub, he makes quite a splash,
And his little barbels droop like a Mexican moustache.

IV Perch

Striped like a chunky peppermint humbug,
He gob-stoppers one of his bag of grayling,
Fencing the rest with a threatening shrug
Of a back-fin spiky as an iron railing.

Ted Walker

Penguins

Penguins dress in curious clothes,
are at home among the bleakest snows.
They throng like well-dined
portly gentlemen with toes
turned in, lined
up for taxis, buttoning
black overcoats,
white scarves dangling
from their throats.

Every penguin expects his
neighbour to be polite.
They don't do things by half.
If anyone transgresses,
they fight
or start wrangling,
their voices loud and hoarse,
their penguin sentences
just a trifle coarse.

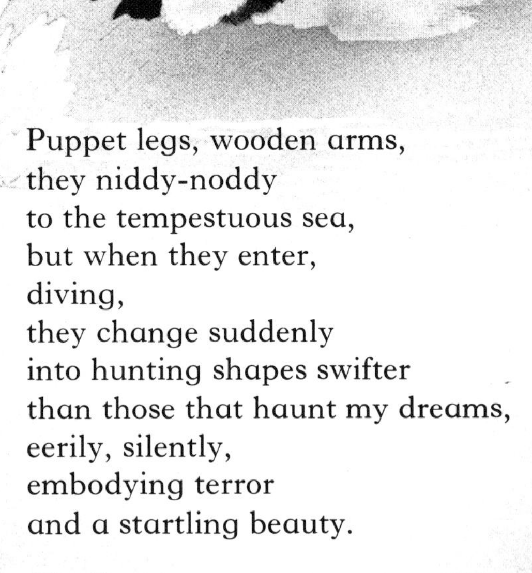

Puppet legs, wooden arms,
they niddy-noddy
to the tempestuous sea,
but when they enter,
diving,
they change suddenly
into hunting shapes swifter
than those that haunt my dreams,
eerily, silently,
embodying terror
and a startling beauty.

Albert Rowe

The Heron

I said to the heron, 'Why do you stand
In that swift-flowing stream in the pebbles and sand
On only one foot?
I'd have thought it would be more convenient to put
Both feet in the stream while you patiently seek
The silvery fish to spear with your beak?'

The heron glared back and his voice quickly rose,
'I'd have thought it was something that everyone knows:
In a warm, feathered hollow one foot I now hold
Because swift-flowing streams are excessively cold.'

Gregory Harrison

A Souvenir

On my Auntie Mabel's mantelpiece.
There sits a seal. Not live and real
Of course, but tiny, dead and real.
Know what I mean? It's made of seal;
Dead thing, but true to life in each detail.

I wonder how a seal must think
(If think it can) or feel at brink
Of sea to hear the culling call
Of man to man as men wade in
To beat with staves this solid flesh
(Still frail) to make a souvenir of Canada
That sits on Auntie Mabel's mantelpiece

John Kitching

The Carver

I look at tree and stone
And whisper: 'Who are you?
Who hides in you?' My knife
In hand, I wait for wolf
Or seal or child or leaf
To guide my patient hand
And free the secret hidden in the bone.

John Kitching

Windy Tree

Think of the muscles
a tall tree grows
in its leg, in its foot,
in its wide-spread toes –
not to tip over
and fall on its nose
when a wild wind hustles
and tussles and blows.

Aileen Fisher

Thunder and Lightning

Blood punches through every vein
As lightning strips the windowpane.

Under its flashing whip, a white
Village leaps to light.

On tubs of thunder, fists of rain
Slog it out of sight again.

Blood punches the heart with fright
As rain belts the village night.

James Kirkup

Pylons

The pylons have stopped
dancing in the field
for one has fallen:

seventy-odd feet
of knitted metal.
in tangled ruin.

Only when you can
walk the length that is
do you realize

the height that has been:
the others still seem
so many dancers

trailing their streamers
in the dust while a
sister ties her shoe.

Keith Bosley

The Scarecrow

The farmer carried some bits of coal,
A pebble or two to try for size;
In the back of the scarecrow's head he made
A mouth, a nose and a pair of eyes.

'There now,' he said, 'that'll fix you up.'
The scarecrow smiled and a scouting rook
Did a fluttery hop and a waddly walk
And croaked to his friends, 'You up there, look.'

And all the rooks in the top of the tree
Started to talk, talk, talk, talk, talk.
And the boss rook said, 'Just watch yourselves;
That man in the field's got eyes like a hawk.'

'Keep a look out, for his coal-black gaze
Can sweep the cornfield now both ways;
Just watch out,' he squawked as he fled,
'That man's got eyes in the back of his head.'

Gregory Harrison

Falling Leaves

The leaves in autumn
Swing on the boughs
Pushed by the wind,
Backwards and forwards
Far above the ground.
The wind blows hard
And the leaves let go
Their hold,
Flying over the wood
Like a flock of birds
And brown as sparrows
Flying above the house.
The wind lets them fall
Helter skelter
Towards the ground
Or sets them spinning
On a roundabout;
You can watch them
And try to catch them
Coming down.
They crowd together
By walls at corners
Or chase each other
Up the road.
Sometimes when you open
The kitchen door
A leaf blows in
And lies exhausted
On the floor.

Stanley Cook

Autumn Song

There came a day that caught the summer
Wrung its neck
Plucked it
And ate it.

Now what shall I do with the trees?
The day said, the day said.
Strip them bare, strip them bare.
Let's see what is really there.

And what shall I do with the sun?
The day said, the day said.
Roll him away till he's cold and small.
He'll come back rested if he comes back at all.

And what shall I do with the birds?
The day said, the day said.
The birds I've frightened, let them flit,
I'll hang out pork for the brave tomtit.

And what shall I do with the seed?
The day said, the day said.
Bury it deep, see what it's worth.
See if it can stand the earth.

What shall I do with the people?
The day said, the day said.
Stuff them with apple and blackberry pie –
They'll love me then till the day they die.

There came this day and he was autumn.
His mouth was wide
And red as a sunset.
His tail was an icicle.

Ted Hughes

Hallowe'en

Witch's fiddle, turnip middle,
Scoop it all out with a spoon.
Curve mouth and eyes
With a careful knife
Beneath a Hallowe'en moon.

Witch-broom handle, long wax candle,
Stick spell-firm in the hole.
Find a match,
Step back and watch
Hushed as a Hallowe'en mole.

Witch-keen sight, strike bright light,
Match to the greasy wick.
See faint flame
Flick and falter,
Rise and stutter.
Part of the Hallowe'en game.

Witch-black cat; put turnip hat
Gently back on the top.
Turn out all moon.
Watch yellow eyes, mouth's flamed rays.
Hark for a Hallowe'en tune.

For the witch's fiddle
And the witch's cat
And the crack
Of a witch-broom handle
Sing a haggard song
On a moonless night
To a turnip lantern candle.

John Kitching

My Christmas List

A police car
A helicopter
A gun that goes pop
A Frisbee
A ball
An Action Man that won't stop
A torch
A guitar
A printing set with ink
A bouncer
A new bear
A submarine that won't sink
A sword
A typewriter
A stove so I can cook
A radio
A Wendy house
Another dinosaur book –
Of course, Father Christmas, it's clearly understood
That I'll only get *all* of this if I'm *specially* good.

Gyles Brandreth

The Mutinous Jack-in-the-box

Why should I always jump up
when they press that stupid button
always at their call?
Next time . . . *next* time . . .
I'll stick my tongue out at them!
I'll spit!
I'll shout rude words!
I'll pull horrid faces!
I'll make their baby cry
and turn their milk sour;
I will!
Just you wait and see!
They can't fool me;
not for ever a slave;
next time I'll be brave.

B o o o o o o o o o o o i i i i i i i i n n n n n n n n g g g!

Oh dear, I jumped up again,
Grinning as though I had no brain.
But . . . just you wait and see,
Next time I'll do it;
I will . . . I will . . .

John Cunliffe

Winter

On winter mornings in the playground
The boys stand huddled,
Their cold hands doubled
Into trouser pockets.
The air hangs frozen
About the buildings
And the cold is an ache in the blood
And a pain on the tender skin
Beneath finger nails.
The odd shouts
Sound off like struck iron
And the sun
Balances white
Above the boundary wall.
I fumble my bus ticket
Between numb fingers
Into a fag,
Take a drag
And blow white smoke
Into the December air.

Gareth Owen

MW.

The Fate of an Icicle

An unthinking icicle
riding a bicycle
went too close to the sun –

singing a song
a newly-born icicle
went to a place
where he didn't belong –

a newly-born icicle
pedalling a bicycle
thoughtlessly
singing his song of the frost
went too close to the sun
and got lost.

Whatever he felt
as he started to melt
he soon fought free of the flame –

which sent back the bicycle
minus its icicle
propelled by a flower
of solar power.

Alan Sillitoe

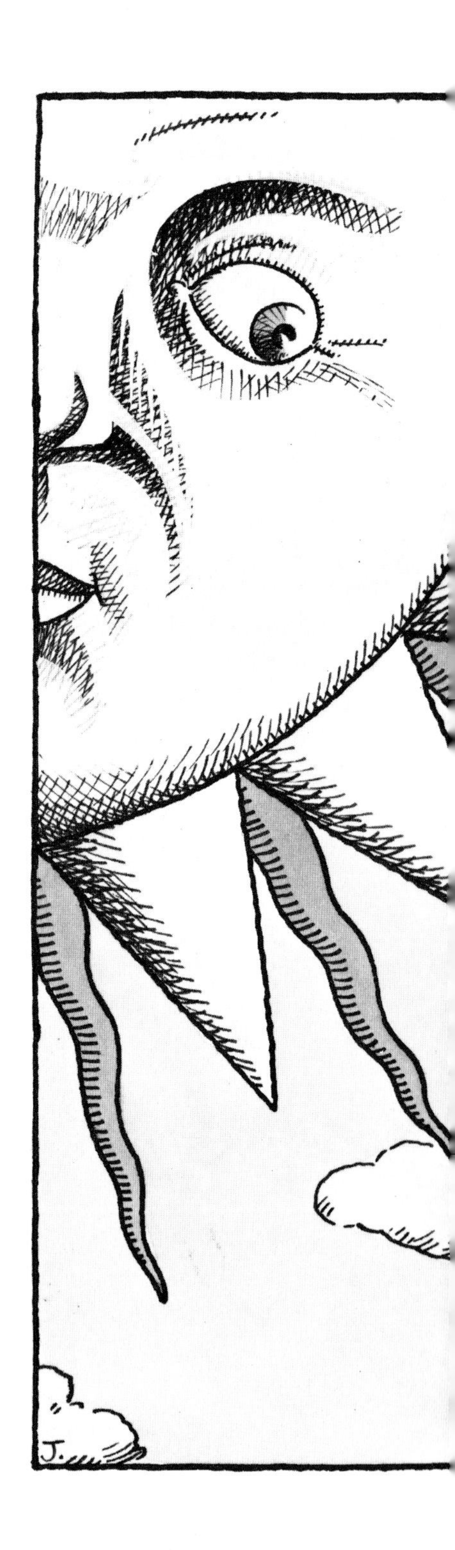

The Kitten in the Falling Snow

The year-old kitten
has never seen snow,
fallen or falling, until now
this late winter afternoon.

He sits with wide eyes
at the firelit window, sees
white things falling
from black trees.

Are they petals, leaves or birds?
They cannot be the cabbage whites
he batted briefly with his paws,
or the puffball seeds in summer grass.

They make no sound, they have no wings
and yet they can whirl and fly around
until they swoop like swallows, and
disappear into the ground.

'Where do they go?' he questions,
with eyes ablaze, following their flight
into black stone. So I put him
out into the yard, to make their acquaintance.

He has to look up at them: when one
blanches his coral nose, he sneezes,
and flicks a few from his whiskers, from
his sharpened ear, that picks up silences.

He catches one on a curled-up paw
and licks it quickly, before
its strange milk fades, then sniffs its ghost,
a wetness, while his black coat

shivers with stars of flickering frost.
He shivers at something else that makes his thin
tail swish, his fur stand on end! 'What's this? . . . '
Then he suddenly scoots in to safety

and sits again with wide eyes
at the firelit window, sees
white things falling
from black trees.

James Kirkup

Snowball Wind

The wind was throwing snowballs.
It plucked them from the trees
and tossed them all around the woods
as boldly as you please.

I ducked beneath the spruces
which didn't help a speck;
the wind kept throwing snowballs
and threw one down my neck.

Aileen Fisher

The New Year

From a surly
Night of winter
Moaning dourly
Over the snows,

I climbed early
New Year's morning
As the pearly
Frost-light rose,

And from a burly
Iron hilltop
Where the dreary
Jackdaws froze,

I saw clearly
Ploughlines where the
Summer barley
Always grows.

Ted Walker

River Winding

Rain falling, what things do you grow?
Snow melting, where do you go?
Wind blowing, what trees do you know?
River winding, where do you flow?

Charlotte Zolotow

Reading

When I play hide-and-seek
In winter books, what do I see?
Do I seek the folk in books?
Or do they quietly hunt
What I have lost in me?

John Kitching

Growing Up

I know a lad called Billy
Who goes along with me
He plays this game
Where he uses my name
And makes people think that he's me.

Don't ever mess with Billy
He's a vicious sort of bloke
He'll give you a clout
For saying nowt
And thump you for a joke.

My family can't stand Billy
Can't bear him round the place
He won't eat his food
He's always rude
And wears scowls all over his face.

No one can ever break Billy
He's got this look in his eye
That seems to say
You can whale me all day
But you'll not make Billy cry.

He has a crazy face has Billy
Eyes that look but can't see
A mouth like a latch
Ears that don't match
And a space where his brains should be.

Mad Billy left one morning
Crept away without being seen
Left his body for me
That fits perfectly
And a calm where his madness had been.

Gareth Owen

Looking Down on Roofs

When I lived in a basement,
The house above my head
Was like a mountain of brick and wood,
A place of weight and dread.
But now I'm in a Tower Block
I can look right down my nose
At the mingy, stingy streets below,
That lie beneath my toes.

When I lived in a basement,
I smelled the damp all night;
And the cats and rats of the neighbourhood
Would choose our yard to fight.
But now I'm in a Tower Block,
It's clean, and fresh, and high, –
And I love to look down on the roofs and know
That I'm nearer to the sky.

Marian Lines

I used to have a little red alarm clock

I used to have a little red alarm clock.
It was my dad's.
He gave me it
and I used to keep it by the side of my bed.

It was very small and it had legs
only the legs were like little marbles –
little metal marbles,
and you could unscrew them
out of the bottom of that little red clock.

One morning
I was lying in bed
and I was fiddling with my clock
and I unscrewed one of those
little marble-leg things
and, do you know what I did?
I slipped it into my mouth – to suck,
like a gob-stopper.

Well it was sitting there,
underneath my tongue
when I rolled over
and – ghulkh – I swallowed it:
the leg off my clock.
It had gone. It was inside me. A piece of metal.

I looked at the clock.
It was leaning over on its side.
I stood it up and of course it fell over.

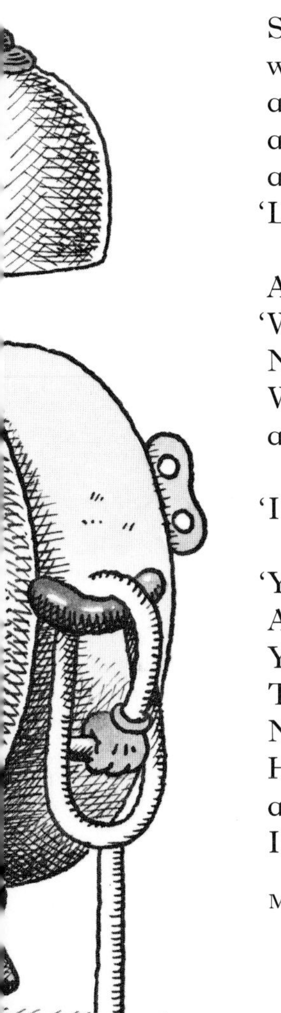

So I got up
went downstairs with it
and I was holding it out in front of me
and I walked in to the kitchen
and I said:
'Look, the clock. The leg. The leg. The clock – er . . .'

And my dad took if off me and he said,
'What's up, lad? Did you lose it?
Not to worry, it can't have gone far.
We'll find it,
and we can screw it back on here, look.'

'I swallowed it,' I said.

'You swallowed it? You swallowed it?
Are you mad? Are you stark staring mad?
You've ruined a perfectly good clock.
That was a good clock, that was. Idiot.
Now what's the use of a clock that won't stand up?'
He held it out in front of him,
and he stared at it. I looked at it too.
I was wondering what was happening to the leg.

Michael Rosen

Chester's Undoing

Chester Lester Kirkenby Dale
Caught his sweater on a nail.
As Chester Lester started to travel
So his sweater began to unravel.
A great long trail of crinkly wool
Followed Chester down to school.
Then his ears unravelled!
His neck and his nose!
Chester undid from his head
To his toes.
Chester's undone, one un-purl, two un-plain,
Who's got the pattern to knit him again?

Julie Holder

My Sock

My sock
is
round my foot
My sock
is
in my shoe.

How can it
not only be
in two places
but 2 shapes?

Socks
are more cunning
than
they let on.

Ivor Cutler

A Little Girl I Hate

I saw a little girl I hate
And kicked her with my toes.
She turned
And smiled
And KISSED me!
Then she punched me on the nose.

Arnold Spilka

I used to watch TV

I used to watch TV
And got the Muppet habit
My brother told me to read this book –
And I got the Hobbit habit.

At school Mrs Lowe read this other book
And I saw the film and liked the song
Now I've got the bright-eyed rabbit habit.

If I could, should I stop it?

John Kitching

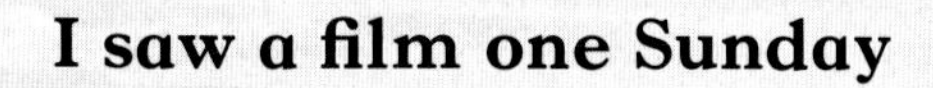

I saw a film one Sunday

I saw a film one Sunday
Called 'Oh! What a Lovely War.'
I couldn't understand it –
But what were they fighting for?

Great Grandad says he was 'in it' –
Fighting the war, I mean.
But he doesn't know what he fought for,
And he just wouldn't say what he'd seen.

John Kitching

I know a man who's got a pebble

I know a man who's got a pebble.
He found it and he sucked it
during the war
He found it and he sucked it
when they ran out of water
He found it and he sucked it
when they were dying for a drink.
and he sucked it and he sucked it
for days and days and days.

I know a man who's got a pebble
and he keeps it in his drawer

It's small and brown – nothing much to look at
but I think of the things he thinks
when he sees it:
how he found it
how he sucked it
how he nearly died for water to drink.

A small brown pebble
tucked under his tongue
and he keeps it in his drawer
to look at now and then.

Michael Rosen

Little Old Man

Little old man hunched and grey
I know you were young once – like me.
But it's hard to believe I'll ever be
the way I see you are today
hunched and grey
little old man
(once young like me).

Charlotte Zolotow

Clothes

My mother keeps on telling me
When she was in her teens
She wore quite different clothes from mine
And hadn't heard of jeans,

T-shirts, no hats, and dresses that
Reach far above our knees.
I laughed at first and then I thought
One day my kids will tease

And scoff at what *I'm* wearing now.
What will *their* fashions be?
I'd give an awful lot to know,
To look ahead and see.

Girls dressed like girls perhaps once more
And boys no longer half
Resembling us. Oh, what's in store
To make *our* children laugh?

Elizabeth Jennings

The Traveller

Old man, old man, sitting on the stile,
Your boots are worn, your clothes are torn,
  Tell us why you smile.

Children, children, what silly things you are!
My boots are worn and my clothes are torn
  Because I've walked so far.

Old man, old man, where have you walked from?
Your legs are bent, your breath is spent –
  Which way did you come?

Children, children, when you're old and lame,
When your legs are bent and your breath is spent
  You'll know the way I came.

Old man, old man, have you far to go
Without a friend to your journey's end,
  And why are you so slow?

Children, children, I do the best I may:
I meet a friend at my journey's end
  With whom you'll meet some day.

Old man, old man, sitting on the stile,
How do you know which way to go,
  And why is it you smile?

Children, children, butter should be spread,
Floors should be swept and promises kept –
  And you should be in bed!

Raymond Wilson

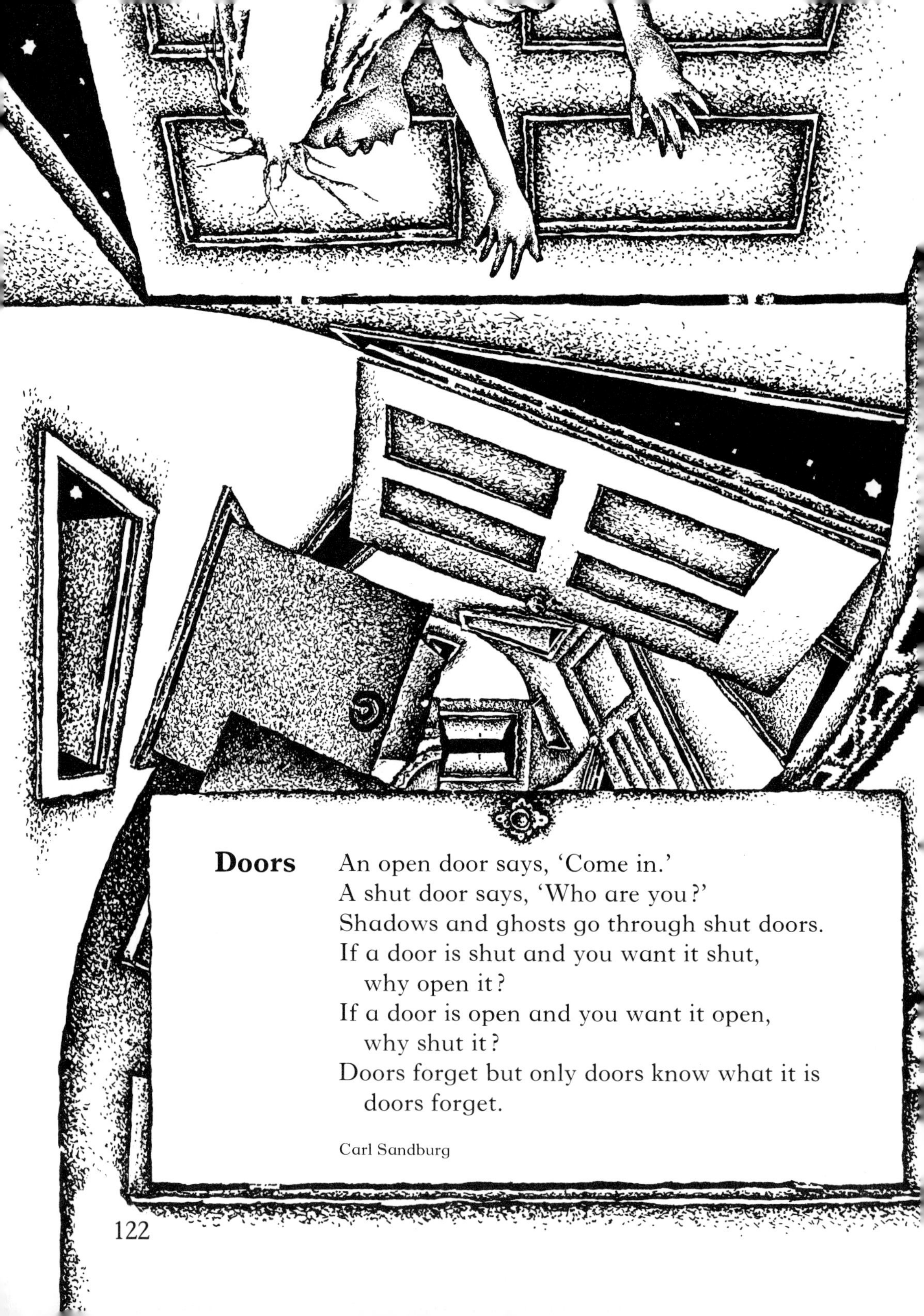

Doors

An open door says, 'Come in.'
A shut door says, 'Who are you?'
Shadows and ghosts go through shut doors.
If a door is shut and you want it shut,
 why open it?
If a door is open and you want it open,
 why shut it?
Doors forget but only doors know what it is
 doors forget.

Carl Sandburg

Why?

I wonder why the brown seed breaks
Under the earth, what hand awakes
The robin in its shell, and brings
Out of the silence song and wings?

Wise men explain away the world
By movement, water, gas, and dust,
Explain how leaves and limbs uncurled
Through dancing days of sunlight must
Turn to the cold, and rust.

But who can tell us, who explain
The reason why we're born to grow
Rich in sorrow, strong through pain?
Life is wiser than we know.

Phoebe Hesketh

Index of first lines

Acknowledgements

The Editor and Publisher wish to thank the following for permission to reprint copyright poems in this anthology. Although every effort has been made to contact the owners of the copyright in poems published here, a few have been impossible to trace. If they contact the Publisher, correct acknowledgement will be made in future editions.

Carey Blyton: From *Bananas in Pyjamas* (1972). Reprinted by permission of Faber & Faber Ltd. Keith Bosley: From *And I Dance* (1972). Reprinted by permission of Angus & Robertson (UK) Ltd. Palmer Brown: From *Beyond the Pawpaw Trees* (Harper & Row Publishers Inc., 1954). William Cole: From *A Boy Named Mary Jane* (1977). Reprinted by permission of Franklin Watts Publishers (London and New York). Charles Connell: From *Versicles and Limericks* (1979). Reprinted by permission of The Hamlyn Publishing Group Limited. Stanley Cook: 'Chips' from *Come Along* (1978, published by the author, Dept. English Studies, Huddersfield Polytechnic); 'Action Man', 'In The Playground' and 'Falling Leaves' are from *Word Houses* (1979, also published by the author). Reprinted by permission of the author. Ivor Cutler: From *Many Flies Have Feathers* (Trigram Press). Reprinted by permission of the author. Roald Dahl: From *James and the Giant Peach* (1967). Reprinted by permission of Allen & Unwin (Publishers) Ltd., and Alfred A. Knopf, Inc. Aileen Fisher: 'Windy Tree' and 'Snowball Wind' from *In The Wood, In The Meadow, In The Sky* (Charles Scribners' Sons 1965). Reprinted by permission of the author. 'Upside Down' from *Up The Windy Hill* (1953). Reprinted by permission of Scott, Foresman and Company. Roy Fuller: From *Poor Roy* (1977). Reprinted by permission of Andre Deutsch. Esther Valck Georges: From *Bits and Pieces* (edited by Peggy Blakely, 1970). Reprinted by permission of A. & C. Black. Robert Graves: From *A Golden Land* (Longman, 1958). Reprinted by permission of A. P. Watt Ltd. on behalf of the author. Phoebe Hesketh: From *A Song of Sunlight* (1974). Reprinted by permission of the author and Chatto & Windus. Russell Hoban: From *The Pedalling Man and other Poems*, Copyright © 1968 by Russell Hoban. Reprinted by permission of Grosset & Dunlap, Inc. and World's Work Ltd. Ted Hughes: From *Season Songs* (1976), Copyright © 1968, 1973, 1975 by Ted Hughes. Reprinted by permission of Faber & Faber Ltd. and Viking Penguin Inc. Elizabeth Jennings: From *The Secret Brother* (Macmillan, 1966). Reprinted by permission of David Higham Associates. X. J. Kennedy: From *One Winter Night In August and Other Nonsense Jingles* (Atheneum, 1975. A Margaret K. McElderry Book), Copyright © 1975, 1977, 1978, 1979 by X. J. Kennedy. Reprinted by permission of Curtis Brown Ltd. and Atheneum Publishers. James Kirkup: From *Shepherding Winds* (Blackie, 1969). Reprinted by permission of Dr. Jan Van Loewen Ltd. Marian Lines: From *Tower Blocks* (1975). Reprinted by permission of Franklin Watts Ltd. (London). Celia McMaster: From *The Blue Peter Book of Limericks* (Piccolo/B.B.C., 1972, edited by B. Baxter & R. Gill). Reprinted by permission of the author. Roger McGough: From *Sporting Relations* (Eyre Methuen, 1974), Copyright © 1974 by Roger McGough. Reprinted by permission of Hope Leresche & Sayle. Samuel Menashe: From *The Many Named Beloved* (Gollancz, 1961) and also from *To Open* (Viking Press, 1974), Copyright © 1974 by Samuel Menashe. Reprinted by permission of the author and Viking Press Inc. Spike Milligan: From *Silly Verse For Kids* (1963). Reprinted by permission of Dobson Books Ltd. Lilian Moore: From *I Feel The Same Way* (Atheneum, 1967). Reprinted by permission of the author. Ogden Nash: From *Custard and Co* (Kestrel Books, 1979) and also from *Verses From 1929 On* (Little, 1959), Copyright © 1940 by the Curtis Publishing Company. This poem first appeared in the *Saturday Evening Post*, 1940. Reprinted by permission of A. P. Watt Ltd. on behalf of the estate of Ogden Nash; and Little, Brown and Company. Gareth Owen: From *Salford Road* (Kestrel Books, 1979), Copyright © 1971, 1974, 1976, 1979 by Gareth Owen. Reprinted by permission of Penguin Books Ltd. Jack Prelutsky: 'The Ogre' and 'The Ghoul' from *Nightmares: Poems to Trouble Your Sleep*, Text copyright © 1976 by Jack Prelutsky. Reprinted by permission of Adam and Charles Black Publ. and Greenwillow Books (A Division of William Morrow & Co.). 'The Hippopotamus' from *Toucans Two and Other Poems* (Macmillan Publishing Company, 1970) and also in *Zoo Doings and Other Poems* (Hamish Hamilton, 1971), Copyright © 1967, 1970 by Jack Prelutsky. Reprinted by permission of Macmillan Publishing Co. Inc. and Hamish Hamilton Ltd. Michael Rosen: 'If you don't put your shoes on' from *Mind Your Own Business* (1974). Reprinted by permission of Andre Deutsch. 'I used to have a little red alarm clock' and 'You tell me', both from *You Tell Me* by Roger McGough and Michael Rosen (Kestrel Books, 1979), Copyright © Michael Rosen, 1979. This collection copyright © 1979 by Penguin Books Ltd. Reprinted by permission of Penguin Books Ltd. Carl Sandburg: From *The Complete Poems of Carl Sandburg* (rev. ed., 1970), Copyright © 1950 by Carl Sandburg; renewed 1978 by Margaret Sandburg, Helga Sandburg

Crile, and Janet Sandburg. Reprinted by permission of Harcourt Brace Jovanovich, Inc. William Jay Smith: From *Laughing Time* (Little, Brown, 1955), Copyright © 1955 by William Jay Smith. Reprinted by permission of the author. Arnold Spilka: From *A Rumbudgin of Nonsense* (Scribners, 1970), Copyright © 1970 by Arnold Spilka. Reprinted by permission of the Frances Schwartz Lit. Agency. Colin West: From *Out of the Blue From Nowhere* (1976). Reprinted by permission of Dobson Books Ltd. Raymond Wilson: From *Time's Delights* (1977). Reprinted by permission of The Hamlyn Publishing Group Ltd. Kit Wright: From *Rabbitting On* (Fontana Lions, 1978). Reprinted by permission of the publisher. Charlotte Zolotow: From *River Winding* (1970). Reprinted by permission of Abelard-Schuman Ltd.

The following poems are appearing for the first time in this anthology and are reprinted by permission of the author unless stated:
Alan Bold: 'The Malfeasance', Copyright © 1980 by Alan Bold. Gyles Brandreth: 'Why', 'Ode to a Goldfish' and 'My Christmas List', all Copyright © 1980 by Gyles Brandreth. Leonard Clark: 'Town Fox', Copyright © 1980 by Leonard Clark. John Cunliffe: 'Orders of the Day' and 'The Mutinous Jack-in-a-box', both Copyright © 1980 by John Cunliffe. Gregory Harrison: 'Legging the Tunnel', 'The Heron' and 'The Scarecrow', all Copyright © 1980 by Gregory Harrison. Julie Holder: 'Shut that door!' and 'Chester's Undoing', both Copyright © 1980 by Julie Holder. Ted Hughes: 'Black Bear' and 'The Musk-ox', both Copyright © 1980 by Ted Hughes. Reprinted by permission of Olwyn Hughes, Literary Agent. James Kirkup: 'The Kitten in the Falling Snow', Copyright © 1980 by James Kirkup. Reprinted by permission of Dr. Jan Van Loewen Ltd. John Kitching: 'Why', 'My goodness! What a big boy you are!', 'Don't spray the fly, Dad', 'My Gerbil', 'A Souvenir', 'The Carver', 'Hallowe'en', 'Reading', 'I used to watch TV', 'I saw a film one sunday', all Copyright © 1980 by John Kitching. Gareth Owen: 'Growing Up', Copyright © 1980 by Gareth Owen. Michael Rosen: 'Juster and Waiter', 'The Longest Journey in the World', 'One Tuesday when I was about ten', and 'I know a man who's got a pebble', all Copyright © 1980 by Michael Rosen. Albert Rowe: 'The Donkey boys' and 'Penguins', both Copyright © 1980 by Albert Rowe. Ian Serraillier: 'The Visitor' and 'The Flea and the Ox', both Copyright © 1980 by Ian Serraillier. Alan Sillitoe: 'The Fate of an Icicle', Copyright © 1980 by Alan Sillitoe. Derek Stuart: 'Elastic Jones', 'Giant Denny', 'My Gramp', all Copyright © 1980 by Derek Stuart. Ted Walker: 'Four Fish' and 'The New Year', both Copyright © 1980 by Ted Walker. Reprinted by permission of David Higham Assoc. Ltd.